SCIENTIFIC AMERICAN | EDUCATIONAL PUBLISHING

SCIENTIFIC AMERICAN INVESTIGATES CAREERS IN SCIENCE

ZOOLOGIST

BY ROSIE BANKS

Published in 2025 by The Rosen Publishing Group
in association with Scientific American Educational Publishing
2544 Clinton Street, Buffalo NY 14224

Library of Congress Cataloging-in-Publication Data
Names: Banks, Rosie, 1978- author.
Title: Zoologist / Rosie Banks.
Description: Buffalo : Scientific American Educational Publishing, an imprint of Rosen Publishing, 2025. | Series: Scientific american investigates careers in science | Includes index.
Identifiers: LCCN 2024037635 (print) | LCCN 2024037636 (ebook) | ISBN 9781725352445 (library binding) | ISBN 9781725352438 (paperback) | ISBN 9781725352452 (ebook)
Subjects: LCSH: Zoologists–Juvenile literature.
Classification: LCC QL50.5 .B36 2025 (print) | LCC QL50.5 (ebook) | DDC 590.92 [B]–dc23/eng/20240814
LC record available at https://lccn.loc.gov/2024037635
LC ebook record available at https://lccn.loc.gov/2024037636

Designer: Leslie Taylor
Editor: Therese Shea

Portions of this work were originally authored by Blair Belton and published as *Be an Zoologist*. All new material in this edition is authored by Rosie Banks.

Photo credits: Cover (main) AzmanJaka/iStockphoto.com; series art (background) jijomathaidesigners/Shutterstock.com; p. 5 James Kirkikis/Shutterstock.com; p. 7 MikhailSh/Shutterstock.com, (inset) Sarunyu L/Shutterstock.com; p. 8 mutinamatyas_photo/Shutterstock.com; p. 9 MriyaWildlife/iStockphoto.com, (dog) Chase D'animulls/Shutterstock.com, (cat) Svetlana Rey/Shutterstock.com, (flea) Vera Larina/Shutterstock.com, (hummingbird) Pixfinity Studio/Shutterstock.com, (slug) Eric Isselee/Shutterstock.com, (polar bear) RilakkuMaxx/Shutterstock.com, (bat) illupic/Shutterstock.com, (octopus) pan demin/Shutterstock.com, (reindeer) Potapov Alexander/Shutterstock.com; p. 11 In2dodoo/Shutterstock.com, (inset) Focus Pix/Shutterstock.com; p. 13 Leandro Avelar/File:Bothrops jararaca.jpg_commons.wikimedia.org; p. 15 ugurhan/iStockphoto.com; p. 16 (fox) BlueBarronPhoto/Shutterstock.com, (eagle) Al Carrera/Shutterstock.com; p. 17 Flatfeet/Shutterstock.com; p. 19 (top) U.S. Air Force photo/File:Powered pooter.jpg_commons.wikimedia.org, (bottom) Gabriela Insuratelu/Alamy.com; p. 21 Brian Gratwicke/File:Ecnomiohyla rabborum.jpg_commons.wikimedia.org; p. 23 Joost van Uffelen/Shutterstock.com; p. 24 Christian Sardet/CNRS/Tara expeditions/File:Plankton species diversity.jpg_commons.wikimedia.org; p. 25 Amri1984/Shutterstock.com, (inset) Sarah_lewis/Shutterstock.com; p. 26 Miroslav Halama/Shutterstock.com; p. 27 Wonderful Nature/Shutterstock.com; p. 29 Sophon Nawit/Shutterstock.com.

Printed in the United States of America

CPSIA compliance information: Batch #CWSA25. For Further Information contact Rosen Publishing at 1-800-237-9932.

CONTENTS

Words in the glossary appear in **bold** type the first time they are used in the text.

THE STUDY OF ANIMALS

Do you love animals? Do you like to watch them to learn how they act? Some people do this as a career! Zoologists are scientists who study animals. This study includes how animals eat, move, and play. It also includes how they **interact** with each other and their habitat, which is the name for their home in nature. Zoologists also compare animals, to find their similarities and differences. This is how they sometimes can spot entirely new species, or kinds, of animals.

Zoologists use their knowledge to keep animal species and their habitats safe. Read on to find out if this is the career for you.

Because there's so much to learn, most zoologists specialize, or focus, on an animal species or habitat. They may have a different job title because of this specialization.

Who's at the Zoo?

The word "zoologist" might make you think of the zoo. A zoologist may study animals in a zoo, but most study them in the wild. In general, the people who care for the animals in a zoo are called zookeepers. Zookeepers feed and take care of many kinds of animals, but they aren't as focused on **research** as zoologists are.

ANIMAL ANATOMY

Some zoologists examine, or study closely, animals' anatomy. This is the word for the structure of living things, including skeletons, muscles, body systems, and even genes. This study helps zoologists figure out how animals should be classified, or grouped. For example, horses, zebras, and rhinoceroses are grouped together as odd-toed ungulates. This means they're hoofed **mammals** (ungulates) with an odd number of toes.

Why do we need animal classification? By studying animals who are related, zoologists can see how different parts of animals have evolved, or changed over time. This helps us understand more about the changes in life on Earth in general.

We know certain dinosaurs evolved into birds because of similarities between the bones of dinosaurs and birds. This photo shows the fossil of a birdlike dinosaur called *Archaeopteryx*.

FUN FACT

IN 2023, SCIENTISTS IDENTIFIED A NEW SPECIES OF SPIDER IN AUSTRALIA. THEY NAMED IT *VENOMIUS TOMHARDYI*, AFTER THE COMIC BOOK CHARACTER VENOM AND THE ACTOR WHO PLAYED HIM IN THE MOVIES!

ANIMAL PHYSIOLOGY

Physiology is the study of the physical and chemical processes that make animals' bodies work. It includes how animals see, run, fly, breathe, and survive in hot and cold places. For example, a bird's wings use bones, nerves, and muscles to make flight possible. Zoologists research how a bird's bones can be so light and yet so strong. They study bird muscles to figure out how they allow some birds to fly for thousands of miles.

Zoologists may work in a lab with wind tunnels and superfast cameras to photograph birds in flight. Or they may be outside, observing flight in natural, wild conditions.

Zoologists often work outside, or do fieldwork, to observe animals. This ornithologist, or zoologist who specializes in birds, put a band on a bird's leg. He will track the bird's **migration.**

Fun Animal Physiology Facts

Dogs

sweat through their foot pads to cool their bodies.

Cats

make more than 100 sounds, while dogs make about 10.

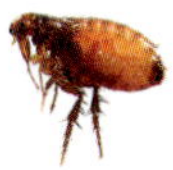

Fleas

can jump 350 times their body length.

Hummingbirds

are the only birds that can fly backward.

Slugs

have four noses.

Polar bear

skin is black.

Bats

are the only mammals that fly.

Octopuses

have three hearts.

Reindeer

eyes turn blue in winter to help them see better.

Zoologists can find out some cool facts about animals as they study them!

WHAT ABOUT BEHAVIOR?

Zoologists know it's not enough to study an animal's body and how it works. Behavior is also important, including what animals eat, where they travel, and how they interact with other animals. This knowledge helps when trying to keep certain animal populations alive.

Sea turtles spend their lives in the deep ocean. Their behavior is hard to observe. However, zoologists know they travel to beaches to lay eggs. Some scientists watch over baby turtles once they hatch. To learn what happens next, zoologists attach tags and **transmitters** to the turtles. This is how they found out leatherback and loggerhead sea turtles travel thousands of miles each year.

As part of an effort to keep baby sea turtles safe, scientists have told people to turn lights off near the coast. The turtles may crawl toward the light and away from the water.

FUN FACT

SOME FEMALE SEA TURTLES FIND THEIR WAY BACK TO THE SAME BEACH WHERE THEY HATCHED YEARS LATER TO LAY EGGS!

ANIMAL ILLNESS

Animals have evolved to survive in challenging conditions. Zoologists study the special ways body systems help creatures survive. Some animals produce poisons to fight off predators. Other animals require little water to stay alive. Knowledge of animal biological processes can help scientists make drugs, foods, and fuels that improve—and even save—people's lives.

For example, birds can carry a deadly disease, or illness, called bird flu. Examining how birds fight bird flu provides possible answers about how to stop the disease from hurting other animals, including people. A disease that animals can pass on to people is called a zoonotic disease.

Captopril is a drug that blocks a substance in the body that causes blood vessels to tighten. It helps treat high blood pressure and prevent heart failure. It was developed from the venom, or poison, of the Brazilian jaraca viper, shown here.

Follow the Instructions

DNA, located in each animal cell, provides the instructions for an animal's body. It controls how it grows, how quickly, and what traits, or features, it will have. Since DNA is **unique** in each body, it can help zoologists compare animals. Animal DNA research may help us understand our own DNA better.

ECOSYSTEM INVESTIGATORS

Animals don't live alone. Many other animals, plants, and other **organisms**, such as bacteria, exist together in an ecosystem. Ecosystems also include the nonliving things in a place, such as weather, soil, water, and sunlight. Zoologists study how animals affect each other in an ecosystem.

In eastern Africa, some zoologists study herbivores, or plant eaters, such as zebras. Others specialize in carnivores, or meat eaters, such as lions. By combining their studies, they can learn how a change in one population, such as a smaller number of zebras because of less grass, can cause a change in another, such as lions going hungry. Zoologists can figure out how best to keep a healthy balance in populations.

Sometimes a team of zoologists keeps track of how a community interacts, not just one or two species. Animals may depend on each other for food, protection, or another need.

Scientists who study whole ecosystems of animals and plants are sometimes called wildlife biologists. They research what land animals eat, how they raise young, and how they affect animals around them. They may work with zoologists who specialize in a certain animal in an ecosystem.

Scientists believe the island fox started to disappear from California's Channel Islands when fish-eating bald eagles died off from pollution and hunting. This opened up nesting space for golden eagles, which hunt island foxes. Wildlife biologists introduced bald eagles back into the Channel Islands. The island fox population went from just 15 around 1998 to 1,800 in 2019.

island fox

golden eagle

In the Netherlands, an area of land called Oostvaardersplassen was turned back into a wilderness populated by horses, cattle, and deer.

An Animal Experiment

At first, the Oostvaardersplassen wildlife area was left alone so it would be completely natural. But a very cold winter, from 2017 to 2018, led to much less vegetation, or plant life. About 60 percent of the large plant-eating animals died. Now, scientists are **intervening** there more, but some think human intervention means this wild area isn't a true wilderness.

BUG SCIENTISTS

Zoology isn't just about big animals. After all, about 80 percent of all animal species in the world are insects! Zoologists who study insects are called entomologists. Some specialize in a single type of insect such as butterflies, ants, grasshoppers, or beetles.

Entomologists may research natural enemies or substances that would stop insects such as locusts from eating valuable crops or termites from eating wood structures. They study honeybees to protect the bees from diseases, so that they can continue to spread pollen and provide honey. Entomologists can also aid in warning and protecting people from diseases carried by insects, such as mosquito-spread malaria and West Nile virus.

Entomologists may use special tools to catch insects without harming them.

FUN FACT

A TOOL CALLED AN ASPIRATOR HAS A TUBE THAT AN ENTOMOLOGIST USES TO SUCK IN AIR AND INSECTS. THE INSECT IS DRAWN INTO A CONTAINER SO IT CAN BE STUDIED. A SCREEN IN THE TUBE STOPS THE INSECT FROM GOING INTO THE ENTOMOLOGIST'S MOUTH!

HELPFUL HERPETOLOGISTS

Herpetology is another branch of zoology. Herpetologists study amphibians, such as frogs, toads, salamanders, and newts; and reptiles, such as snakes, lizards, turtles, and crocodiles. Herpetologists may study the effect of construction near a swamp on amphibians or reptiles that live there. They recommend ways to protect animals if building has a harmful impact.

Herpetologists are researching the disappearance of frogs all over the world. Reasons include disease, pollution, **climate change**, and habitat destruction. Herpetologists are working to find out how to help frogs. Many amphibians are keystone species in their habitats. That means their ecosystem would change a lot if they disappeared.

FUN FACT

AMPHIBIAN SPECIES HAVE EXISTED FOR HUNDREDS OF MILLIONS OF YEARS, SINCE THE TIME OF THE DINOSAURS!

Rabbs' fringe-limbed treefrog is a possibly extinct frog species that had lived in forests of Panama. Herpetologists think as many as 200 frog species have gone extinct since 1970.

ZOOLOGY AT SEA

Some zoologists focus on animals in marine, or ocean and sea, ecosystems. They're called marine zoologists. (Marine biologists study plants too.)

Marine scientists usually work near oceans, lakes, rivers, or other sources of water. Research may mean boating, swimming, and diving to observe and record animals. Common studies include monitoring the impact of pollution on marine animals and searching for ways to save **endangered** species.

Marine zoologists often specialize even further. Malacologists study mollusks, which are animals without a backbone that usually have a shell. Mollusks include oysters, snails, and clams. Mollusks naturally clean water so they're really important to the health of marine ecosystems.

In 2023, harmful **algae** along parts of the Southern California coasts sickened and killed hundreds of sea lions. Marine scientists rescued dozens, holding them in a safe place until the danger passed.

Invasive Species

Marine zoologists also battle invasive species at sea. These are species that aren't native to an area and do great harm when introduced. Zebra mussels are one example. They grow on the sides of ships, clog up water pipes, carry disease, and cause other problems in their new habitats. Scientists have found safe substances to put on ships to discourage mollusks from growing on them.

The oceans are full of life, from huge whales to tiny **plankton**. About 25 percent of all ocean organisms call coral reefs home at some point in their life cycle, even though reefs cover less than 1 percent of the ocean floor.

Reefs are alive—they're made up of tiny living plankton-eating animals called polyps. Unfortunately, perhaps as many as 70 percent of coral reefs have been damaged or destroyed by climate change, pollution, and people's activities. Some marine zoologists specialize in protecting coral reefs from threats since they're so important to the health of the oceans. Their loss also threatens people who depend on sea creatures for food.

FUN FACT

ALL OCEAN ANIMALS DEPEND ON TINY PLANKTON, PICTURED HERE, FOR SURVIVAL, WHETHER THEY EAT IT THEMSELVES OR EAT ORGANISMS THAT EAT IT.

According to a United Nations scientific study, about 14 percent of the world's coral has died since 2009.

dying coral

FISH SCIENTISTS

Zoologists who study fish are called ichthyologists. Fish are essential parts of their habitats as well as a food source for people. Ichthyologists spend a lot of time working on boats. They may count fish in a population to figure out if a species is healthy. They may identify new species too.

Some fish populations have shrunk so much that fishermen aren't allowed to catch that species in certain waters. For example, cod in the North Atlantic Ocean have been protected for many years. Laws don't always help enough, though. In the last 50 years, overfishing has shrunk the world shark and ray populations by more than 70 percent.

cod

Whale sharks are the largest fish, sometimes reaching 40 feet (12 m) long. They're an endangered species.

Farms Too

Farm fishing is a step to stopping overfishing problems. Fish farms are areas of water surrounded by nets or cages where some kinds of fish grow. Ichthyologists may monitor farms to make sure the fish inside are healthy and that they don't have harmful effects on the fish and water outside the nets.

HOW TO GET STARTED

Zoologists usually need a college **degree**. If a zoology major isn't a choice, a similar science major is acceptable to some employers. An even higher degree means more job opportunities. But you don't have to wait until college to start learning. Classes you're taking right now are building blocks for your future studies. Your biology and earth science classes are especially important. Zoologists need to have great reading, writing, and research skills too. After all, a new discovery will need to be shared with the world!

Zoologists can really make a difference in the world—and encourage other people to do the same! Is this the science career for you?

FUN FACT

KEEPING LOCAL ECOSYSTEMS FREE OF TRASH IS A GREAT WAY TO HELP THE ANIMALS IN YOUR COMMUNITY!

Where Zoologists Work

GOVERNMENT AGENCIES
do research and organize efforts to help species in trouble

MUSEUMS
research, organize animal collections, and educate the public

AQUARIUMS
care for marine animals and educate the public

ZOOS
care for animals and educate the public

WILDLIFE AGENCIES
monitor habitats and animal populations

SCHOOLS
educate students about animal life

Zoologists, with their science background, have a wide choice of employers. Some jobs require more experience than others.

Volunteer

Sometimes, young people can **volunteer** at zoos and wildlife parks to get experience with animals. They may even meet and learn from real zoologists. However, zoology can also be practiced at home. Bird-watching, bug collecting, and wildlife observations can be done in local parks and backyards.

GLOSSARY

algae: Plantlike living things that are mostly found in water.

climate change: Long-term change in Earth's climate, caused mainly by human activities such as burning oil and natural gas.

degree: An official paper given to someone who has completed certain classes at a college or university.

endangered: In danger of dying out.

interact: To have an effect on each other.

intervene: To get involved with something to have an effect on an outcome.

mammal: A warm-blooded animal that has a backbone and hair, breathes air, and feeds milk to its young.

migration: Movement to warmer or colder places for a season.

organism: A living thing.

plankton: A tiny plant or animal that floats in the ocean.

research: Studying to find something new.

transmitter: A device that sends out radio waves.

unique: One of a kind.

volunteer: To work without pay.

FOR MORE INFORMATION

Books

Claybourne, Anna. *Zoology*. London, UK: Wayland, 2023.

Howard, Jules. *The Who, What, Why of Zoology: The Incredible Science of the Animal Kingdom*. Beverly, MA: Wide Eyed Editions, 2023.

Websites

Museum of Zoology
www.museum.zoo.cam.ac.uk/learning-outreach/young-zoologists-club
Join a club for young people interested in zoology.

Science Careers: Zoologists
www.sciencekids.co.nz/sciencefacts/careers/zoologist.html
Check out a list of facts to know about this career.

What Is Zoology?
www.amnh.org/explore/ology/zoology
The American Museum of Natural History has games, videos, activities, and more about this fascinating science.

INDEX